EVERYTHING YOU ALWAYS WANTED TO KNOW ABOUT CAMCORDERS
(But Nobody Had the Answers)

A One-on-One, Very Easy-to-Understand Guide
on How to Buy, Operate and Make Money
with a Camcorder

From the author of "Everything You Always Wanted To Know About Video"
(But Nobody Had The Answers)

JOHN P. JOHNSTON

Approved by:

Professional
Videographers
Association
of **A**merica

Special thanks to Irma, Lynn, Mary, Tim, and Lib. Without their love and support none of my writings would have gotten farther than the circular file.

Copyright 1989, Video One Productions,

All Rights Reserved.

1st Printing-July 1989

2nd Printing-October 1989

ISBN 1-877725-01-3

VIDEO ONE PUBLICATIONS
3474 Dromedary Way, #1304
Las Vegas, NV 89115
(702) 643-2880

Re-order information can be found on the final pages of this book.

EVERYTHING YOU ALWAYS WANTED TO KNOW ABOUT CAMCORDERS

A lot of people ask me what type of magazine they should subscribe to that would help them with their video productions. Without a doubt, I always tell them to read Videomaker. It only costs $9.97 for a year (six issues). To subscribe, write: Videomaker Magazine, P. O. Box 3727, Escondido, CA 92025-9645.

Table of Contents

INTRODUCTION	11
WHAT IS A CAMCORDER?	12
BUYING A CAMCORDER	13
BETA	13
VHS-C	13
8mm	14
WHAT TO LOOK FOR IN 8MM CAMCORDERS	15
HIGH RESOLUTION 8MM	16
FULL SIZE VHS CAMCORDERS	17
Price	17
Features	17
S-VHS	18
Digital VHS	18
EASIER BLANK VIDEO TAPES	20
TRADEMARK FRAUD	21
SHOULD I CLEAN MY VIDEO HEADS	22
HOW TO MAKE COPIES	24
Equipment Needed:	24
Price to Charge:	24
Step-by-Step:	24
CAMCORDER ACCESSORIES	26
Video Tripods	27
Character Generators	27
Easy Animation and Titling	27
Gyro Video Lens	28
Hemispheric Lens	28
Telephoto & Wide Angle Lenses	29
The "Must Have" Accessory	29
Eyeopener	30
Extra Battery Packs	31
Something New	31

External Microphones 32
Carrying Cases 33
VIDEO BATTERIES 34
LUX: WHAT IS IT? 35
VIDEO LIGHTING 35
TRACKING CONTROL 36
DirectED PLUS 37
HOW TO MAKE MONEY IN VIDEO 40
STANDARD TAPE SET-UP 42
TRANSFERRING STILLS ONTO
VIDEO TAPE 43
Equipment Needed 43
Price to Charge 43
How to Transfer Photos 44
How to Transfer Slides 45
HOW TO TRANSFER 8MM, SUPER 8MM, OR 16MM FILM
ONTO VIDEO TAPE 47
HOW TO TRANSFER MOVIE FILM ONTO VIDEO 48
HOW TO VIDEOTAPE A HOUSE
FOR INSURANCE PURPOSES 49
Equipment Needed 49
Price to Charge 49
Step-by-Step 49
HOW TO MAKE A VIDEO WILL 51
Equipment Needed: 51
Price to Charge: 51
Step-by-Step: 51
INTERVIEWING FOR MONEY 53
Equipment Needed: 53
Price to Charge: 53
Doing the Interview: 54
HOW TO VIDEOTAPE A WEDDING 55
Equipment Needed 55

IDENT-A-KID	59
Equipment Needed	*59*
Price to Charge	*59*
Step-by-Step	*59*
TRANSFERRING IMPORTANT DOCUMENTS	61
Equipment Needed	*61*
Price to Charge	*61*
Transferring Documents	*61*
BUSINESS VIDEOS	62
If You Use And Send Video Tapes	*62*
HOW TO VIDEOTAPE BIRTHS	63
Equipment Needed	*63*
Price to Charge	*63*
Step-by-Step	*64*
ADDING MUSIC TO VIDEOS	65
HOW TO USE THE ZOOM	67
HOW TO USE THE MACRO LENS	68
HANDS FREE CAMERA SUPPORT	69
DEW INDICATOR	70
TIPS ON FILMING	71
SPECIAL EFFECTS	72
NEWSHOUND	73
GLOSSARY OF VIDEO TERMS	74
ENDORSED AD	30
ENDORSED ADS	77
REORDER INFORMATION	80

Getting more shouldn't mean paying more.

OPTEX MICROPHONES
Built-in microphones are fine but sometimes the high-fidelity sound of an external microphone is needed. Our mikes are designed to give the clearest sound possible.

OPTEX LENSES AND FILTERS
Now you can extend your creativity with a selection of high-quality lenses and filters.

Choose from telephoto and wide angle lenses or experiment with special effects filters such as cross star, multi-image, and polarizers. All lenses and filters are designed to fit most camcorders.

OPTEX VIDEO LIGHTS
We have a wide variety of video lights including cordless models, AC or DC operations, or in combination, to provide just the right lighting for any scene.

You get more with OPTEX Video accessories. More power, more variety at the right price. Creativity is at your finger tips with OPTEX Video accessories. Available where most video camcorders are sold.

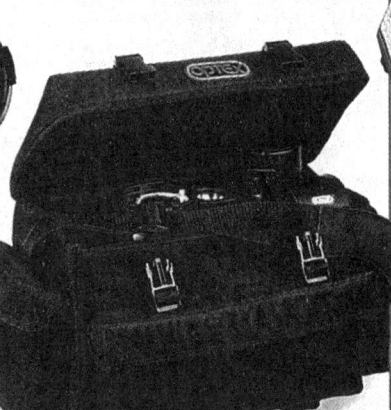

OPTEX BAGS
A good carry-all bag is an important accessory for video recording. OPTEX video bags are constructed of tough durable fabrics to resist tears and soiling. Nylon web straps provide easy portability and ABS plastic pinch clip closures stay secured. And of course there's always enough room to carry extra tapes, spare cables, batteries etc.

OPTEX TRIPODS AND MONOPODS
Each one of our tripods and monopods has been especially engineered for the specific needs of camcorder users. Convenient pan and tilt markings, smooth-action pan heads, and quick release mounts stand apart from the competition.

Sturdy anodized aluminum construction provides solid support. And you will appreciate such added touches as rubber feet and flip lock legs. Available in basic to full-featured models.

OPTEX ON-BOARD BATTERIES
Never miss a shot with our range of on-board rechargeable batteries. Available in 1, 1½ and 2 hour versions to fit most camcorders, these battery packs provide long-lasting power when they're needed most.

OPTEX®

OPTEX INC., ELGIN, ILLINOIS 60123 • IN CANADA: OPTEX CORPORATION, DON MILLS, ONTARIO M3B 2T5

INTRODUCTION

Greetings, it is my intention in this book to give you a no-nonsense, extremely valuable guide to all phases of the camcorder. From the different types to what type is right for you. From the accessories to how to make money in your video productions.

This book is written in English, not Japanese translated into English. (No slur on the Japanese intended. My first wife was Japanese.) But the owner's manuals that come with video equipment nowadays are more techno-oriented than they should be and a lot of the products originate in Japan. Consequently, when the translation from Japanese to English is done they may not say anything you can understand.

In this guide I will try to stay away from using specific brand names, but rather focus on specific features. If you follow the guide of features, you will automatically eliminate a lot of brands or types of camcorder equipment. I have, however, named some brand names of equipment I would fully endorse and have permitted advertising of these superior products in the book.

In this book I will also show you how to take care of, and add extra years of life to, your equipment. I've written this book solely for the consumer. I've broken each subject down so that it is as easy to understand as possible. I've added my personal views as a video expert (not a professional writer) and even some humor. It is my hope that this book will be of help to video enthusiasts and novices alike.

WHAT IS A CAMCORDER?
(Or One Piece Is In; Two Pieces Are Out at the Beach.)

Unlike the separate video camera hooked by an umbilical cord to a video recorder unit of just a few years ago, a Camcorder (video camera and video recorder in one unit) has become the wave of the future. There are four formats of Camcorders: VHS, VHS-C, BETA, 8MM. These four formats are subdivided down even further into: VHS, S-VHS, DIGITAL VHS, S-VHS DIGITAL, VHS-C, S-VHS-C, BETA, SUPER BETA, ED BETA, 8-MM, HI-8MM. They range in weight from about 2.2 lbs to upwards of 12 lbs. (loaded).

The new Camcorders function like the older two-piece units except they probably don't have the tuner/timer function that you would find on your home deck VCR. It's that simple. Also, the Camcorders of today have a lot more automatic features than the two-piece units.

BUYING A CAMCORDER
(What You See is What You Get Section)

There are several types of Camcorders, so let's break them down.

BETA
The same reasons for not buying a BETA VCR apply here. Especially if you go on a trip and want to rent movies to watch in a motel. P.S. if you do watch a movie in a motel be sure to call the desk and tell them what you are doing. Most motel TVs have a switch that trips an alarm in the office if you disconnect the cable line.

VHS-C
We call these Mickey Mouse corders. Advantages are that they are light, usually between 2-4 lbs., and they are small and not too expensive. However, sometimes big problems come in small packages. Some disadvantages are:

a) Most VHS-Cs cannot record and playback on the road.

b) Most only have a viewfinder and do not show a true picture of what you are filming.

c) Most have the viewfinder in the center of the Camcorder (that's good if you are a cyclops. However, not so good for humans).

d) You will need six -- count them, six -- tapes to do what one VHS or 8mm Camcorder video tape does, unless you record in the SLP or six-hour mode. I don't know about you, but I wouldn't want to record Disneyland in the six-hour mode, then make copies of that tape and send it to the family. The quality would be greatly deteriorated.

e) Most do not have a zoom, or they only have a 3-to-1 zoom. All in all the disadvantages definitely outweigh the advantages. They are usually cheaper than the other Camcorders (by about $200). But, you get a lot of Camcorder for the extra $200 if you spend it. I would wait or buy an 8mm (which is fully compatible with VHS) if you want or need a small Camcorder.

What if you bought a VHS-C Camcorder that does not play back in the unit, and you only have one full size VCR deck unit at home. You can take the VHS-C tape and put it into a VHS adaptor and play it in your VCR, but you can't make a copy to send to grandma unless you rent or buy another full size VCR. I have found one person in 20 people actually did buy a VHS-C Camcorder and was happy with it after he bought it.

8mm

The only bad thing going for it, is its name. You will probably recall the 8mm films from the Dark Ages of Movie Cameras. Don't get me wrong, it was good stuff in its time. But, a lot of people still get the two confused -- 8mm movies and 8mm video. 8mm refers to the width of the tape. VHS video tape is 1/2 inch or 16mm.

Almost all 8mm video tape can be played back in the eyepiece. You can run right from the camera to any VHS or BETA to make a copy with only one VCR. The quality is very, very good because the tape used for the 8mm video is five times denser than VHS or BETA. The camera weighs approximately 2 1/2 lbs. on the average. You can record up to two hours on one tape about the size of an audio cassette. The zoom is usually a 6 to 1 zoom which is not too bad.

WHAT TO LOOK FOR IN 8MM CAMCORDERS
(Good Things Come in Small Packages)

1) Auto Focus

2) Six to one zoom or better with a macro. A macro lens will allow you to get to within 3/16 of an inch of the object you are filming while remaining in focus. This is great for putting photos onto video tape or titles.

3) A diopter on the eyepiece. This is great if you wear glasses.

4) An eye piece that you can bring over to your left eye (most people that are left-handed are also left-eyed). With some cameras you need to buy an adapter that will allow you to do this.

5) Ability to add a character generator later (This allows you to add titles to your film.)

6) Almost all 8mm have a flying erase head. Make sure yours does.

7) If you can buy one with video and audio insert editing do so. I will explain more about this in the full size VHS Camcorder section.

8) A quick review button. With this you can review, through your eyepiece, your last 5 seconds of filming and then automatically go back to where you left off. This is good to make sure you got that great shot.

HIGH RESOLUTION 8MM
(Small Things Getting Better)

As with televisions moving aside to Monitor/Receivers, Beta going to ED-Beta, VHS going to S-VHS, 8mm will find its ultimate in High Resolution or High Definition 8mm. Boasting about as many lines of resolution as S-VHS (around 425 to 450 lines or more).

As with every new technological breakthrough it will, like S-VHS have a higher price tag also. But, it does prove that you get what you pay for.

FULL SIZE VHS CAMCORDERS
(Grabbing a Whale By Its Fin)

If you are like most people and are not worried about the size (about twice the size of 8mm or VHS-C) or the weight (average weight is 6-10 lbs.), then this is the Camcorder for you. In my book, there are four types: poor, fair, great, best; which also correspond to money: cheap, medium priced, big bucks, bigger bucks. Does "you get what you pay for" ring a bell?

Let's break it down.

PRICE
Poor - is really not poor at all when you consider price range: $799 to $1,199.

Fair - $1099 - $1,499.

Great - This is the S-VHS (Super VHS) which ranges from $1299 - $1,699.

Best - This is the digital S-VHS from $1499 - $1,999 or higher.

FEATURES
Poor - is really not too bad. They usually have a 6-1 zoom. Try to find one with an 8-1 zoom. Almost all, if not all, have auto-focus, power zoom, diopter (for eyeglass wearers). But the number one flaw is that they do not have video/audio insert editing or a flying erase head. That's why I would recommend going for the Fair Camcorder.

Fair - this is the one I would recommend for 85% of all Camcorder buyers (especially first time buyers). It has all the goodies of the poor ones, but also a flying erase head and video/audio insert editing. Let me explain without getting too technical.

Suppose you filmed some beach shots and later wanted to add some photos at some dead spots and some nice background music. With the poor range of Camcorder you can add photos at the end and forget about the music (unless you added it at the same time you made a copy). The price difference is about $200, but you can do about $1,000 worth of extra things with it. See the sections on how to transfer photo's, slides, letters and film. Also see sections on audio dubbing and how to make money.

Great - S-VHS
These are great. There are two types of Super VHS:

(1) S-VHS-C (If you have to have a VHS-C this is the one to get)

(2) Full size S-VHS. If you want the best picture, have a lot of money, and make a lot of copies (to send to granny) go for it! Make sure that, for this kind of bucks, you get one with video/audio insert editing or it would be like buying a Mercedes to take to a demolition derby. What is so great about S-VHS is that it records over 400 lines of resolution and the color is great, plus you hardly lose quality when you make copies. The only drawback is the tapes usually cost over $10 each (I've seen them at times over $15).

Best - Digital VHS
These are the best: OK. Let's say you have money falling out of your pockets when you sit down, or that you want the absolute best home Camcorder on the planet. This is the one for you! It should have all the usual goodies, plus an 8-1 or 10-1 zoom and unbelievable digital effects. Most have three different wipes. That is where, when you pause the camera it will digitalize the last frame and then when you start the camera up again the new recording will either wipe from right to left over the last frame (see illustration on next page) or come out from the middle or fade the new picture from the old. You can pause/still a shot and then split screen or P-N-P (Picture-N-Picture) in about nine different ways. It also has other types of digital effects (depending on the Camcorder) to make a truly spectacular home production. I have made TV commercials for several companies with

mine. The only drawback is size and weight. The camera is about four inches longer than most full sized VHS Camcorders, so finding a case could be difficult. As for weight, it is about the weight of a good sized baby boy. About 9-12 lbs which means you also need a good sized heavy duty video tripod (see tripod section). Note: you cannot record in S-VHS, remove the tape, and then play it on a conventional VHS (it will look like you came from another planet) but you can transfer dump onto conventional VHS tape the same way as for VHS-C or 8mm. Or, you could record the digital effects in the conventional VHS mode of a digital Camcorder.

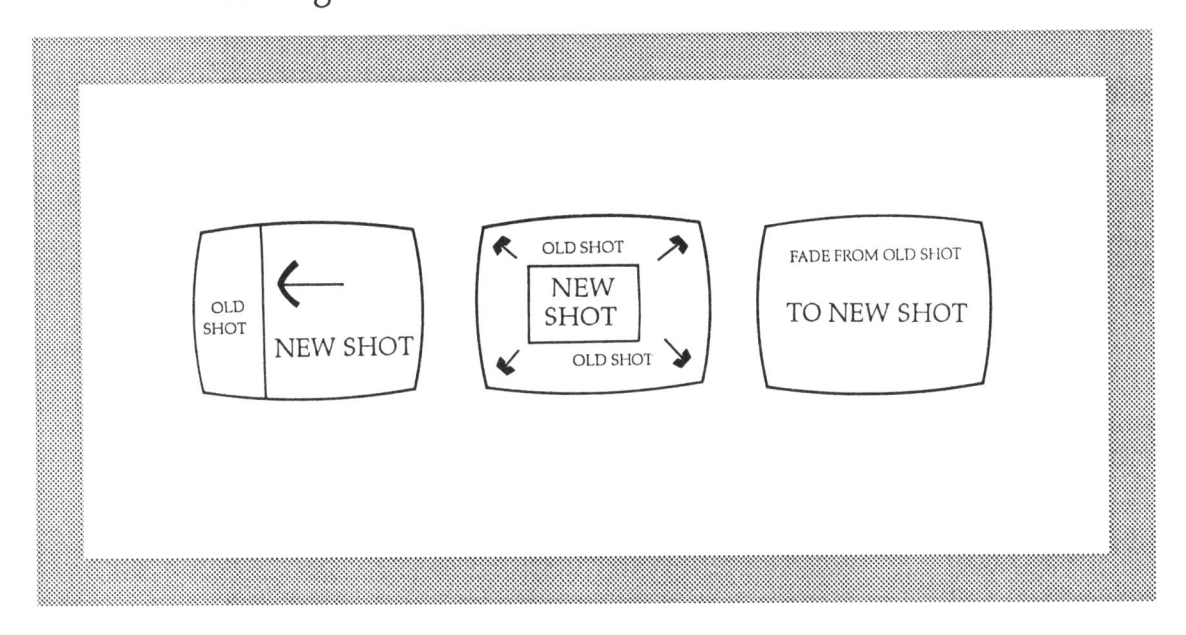

EASIER BLANK VIDEO TAPES

SKC will be releasing a new video tape labeling system, early in 1989, that should eliminate the confusion about which video tapes to buy for what need. They are even broken down by pictures:

Music Video
Grade
A premium quality tape, especially designed for highest quality stereo reproduction, with excellent video quality.

Library
Grade
A better quality tape used to preserve those irreplaceable "keepsake" video recordings for years to come.

Maxplay
Grade
Record and play over and over with longer tape life and better picture quality. Use this grade, or better still, Video Camera Grade, to record Soap Operas or even your favorite TV shows over and over again.

Video Camera
Grade
The professional quality videotape for recording special events and for Camcorder use. It is also best for tape-to-tape copying.

These labels are not an indication of tape quality, just an indication of which types of tapes are best for your particular application. Congratulations, for being one of the first to finally make it easier for the consumer.

TRADEMARK FRAUD
(Look Before You Buy)

Now that I have your attention from reading the title of this section, let me explain.

You may own non-registered video tapes. The tapes that are properly registered will have the following symbol somewhere on the case (this **exact** symbol, not just a close approximation). Some people will try to fool you by putting something that resembles this logo on their non-registered tapes. Having this logo does not mean the tape is the best, but it does mean that it meets certain specifications as to case design, video tape construction, etc. Do not buy any video tape that does not have this exact VHS mark. Let's be more specific. The box around the VHS has to be there. Notice also, the exact way VHS is written. It has to look exactly like this: VHS not VHS or (VHS) . These (what I call) fraud tapes can kill your VCR. Don't be a fool. Look before you buy. Some companies put on the exact same outside case design, but one will have been registered and has the VHS logo while the other does not. The non-registered tapes are almost always very bad in quality. These people are real tricksters. This is not limited to blank video tapes. Some of the prerecorded movies are put on non-registered video tapes. An example of this would be some of the childrens tapes that you see for under $10.

SHOULD I CLEAN MY VIDEO HEADS
(Age Old Question Section)

Of all the questions ever asked about VCRs, this one is on the top of the list. In almost every VCR magazine there is an article about cleaning your heads or not cleaning your heads. On the pro-clean side the experts tend to agree that, because most people buy inexpensive blank video tapes, and they do not examine a rental tape before they rent it (please see the section on renting video tapes), it is necessary to clean their video heads more than once a year. On the con side the experts say for the same reasons, the customer will buy a cheap head cleaner and that all maintenance should be done by a qualified technician. I'm on the pro side, if and only if, you follow this procedure.

1) You do not put your VCR on top of your TV or within 12 inches of the bottom of your TV. It can, over a period of time, damage the VCR or TV. It would be OK to put your VCR on top of most monitors as they should be shielded.

2) Examine rental tapes before you rent them. Read that section.

3) Do not use a cheap head cleaner. I will recommend a head cleaner. Keep in mind that I'm like consumer reports in this book. I do not now or forever more receive money from any company for endorsing their product or products. I would recommend the Geneva (formerly Nortronics) brand of head cleaner. But I would recommend that you use it after 60 to 80 hours of playing and/or recording. Notice that I said playing and recording. If I recorded a two-hour movie, and then two half-hour programs, and then, one night sat down and watched everything that I had recorded, some people would say that I had used my machine only 3 hours. Wrong, I used it for six hours. I'm not saying you should keep a minute-by-

minute log, just keep a mental note of your playing and recording hours.

4) If you follow the first three steps , you should still also have your VCR and video heads cleaned by a professional, but instead of once a year, you can extend the time to once every year and a half. I've encountered hundreds of cases where someone has brought me a VCR that wouldn't work. Put it on the counter for service. When I asked them when was the last time the machine was cleaned, they would smile with naivete. "Oh. I've never cleaned it in three years". Like it was something to be proud of. Unlike TVs or stereos, a VCR *does* need special care. Being a video expert I take this personally. To me it's like running a car off a cliff and then telling a mechanic that there might be something wrong with the front end. On the positive side, if you follow this guide, your VCR will last two to three times longer than the same VCR without this type of care.

5) When getting your VCR heads cleaned, have it done by a professional. Out of 35 video rental places that advertised that they cleaned video heads and machines, I would only recommend two. And they sent their's out to be cleaned. Out of the other 33, not one employed personnel with more than four hours training on VCR repair or cleaning. Not one had a VCR belt in stock or knew how to replace one. Not one knew where or how to replace internal fuses. These are the three most common problems of VCRs. So, ask for training credentials.

6) If you own a Camcorder I would recommend that you buy the **special formula** head cleaner by Geneva. It is specially formulated with a pure TF cleaning solution which I have found to be the best in removing environmental contaminants.

HOW TO MAKE COPIES

Equipment Needed:
Either a Camcorder and a VCR or two VCRs. If both units are HQ equipped, one should have an edit switch. You will also need a set of RCA to RCA cables.

Price to Charge:
$20 for the first copy with case. $15 for each additional copy up to 10. $9.95 each after that. Do not do this for free or for the cost of the tapes alone. Your equipment wear and tear and your time are worth something.

Step-by-Step:
The set-up is quite easy. The illustration on the next page shows two VCRs, one (VCR [a]) as the play unit and the other (VCR [b]) as the recording unit. For a better picture, use RCA to RCA cables. Also, notice the illustration is set up for a stereo or Hi-Fi hookup. If dubbing between two mono VCRs, use only one RCA line. If dubbing from a Stereo VCR to a mono VCR, use a "Y" connector going from the Stereo VCR to the mono VCR to allow both lines from the Stereo of Hi-Fi left and right to be hooked up. If you don't use both lines from the Stereo VCR you will only get part of the sound to transfer across. It would also be helpful to hook up the VHF output of VCR(b) to a TV or hook-up the video and audio outputs of VCR(b) to a monitor/receiver to monitor the recording. You can also hook up more VCRs (called *stringing up*) to make more copies. Do not, however; hook up more than four VCRs this way without using a distribution amp or processing amp. A distribution amp goes in-line between the first VCR and the other VCRs (see illustration). A processing amp hooks up the same way.

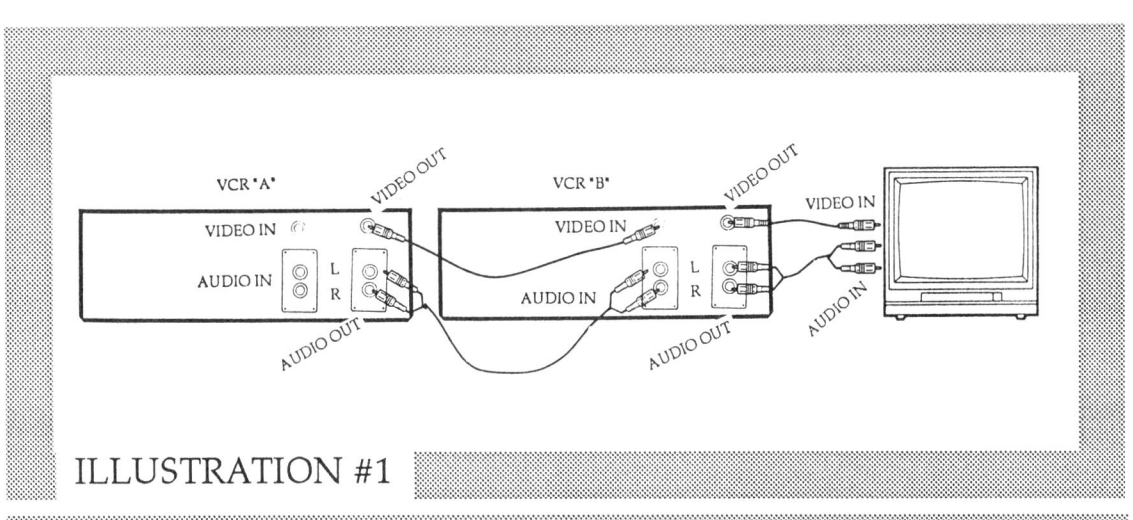

EVERYTHING YOU ALWAYS WANTED TO KNOW ABOUT CAMCORDERS

ILLUSTRATION #1

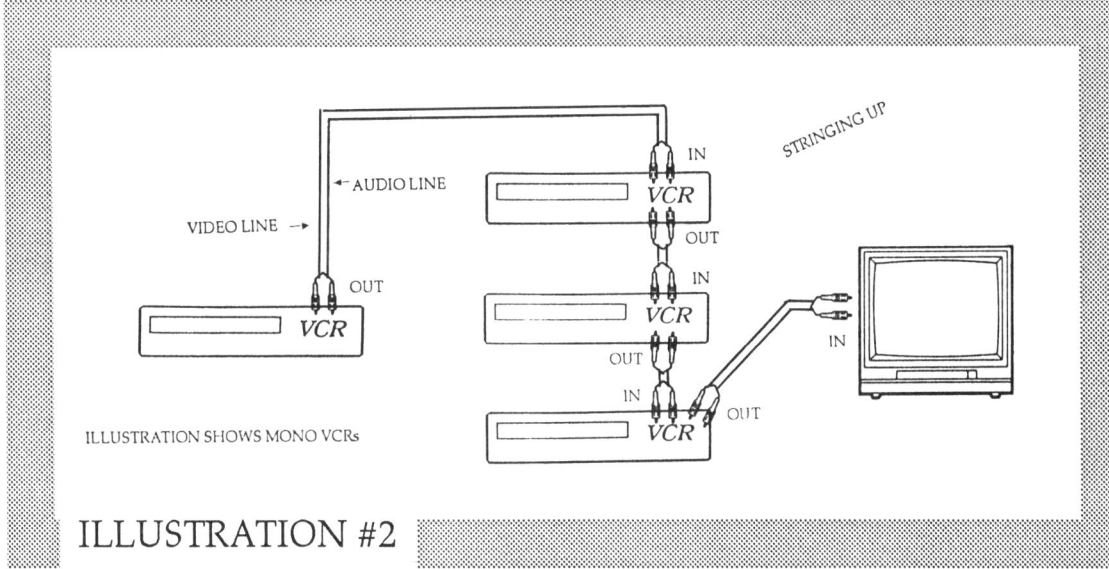

ILLUSTRATION #2

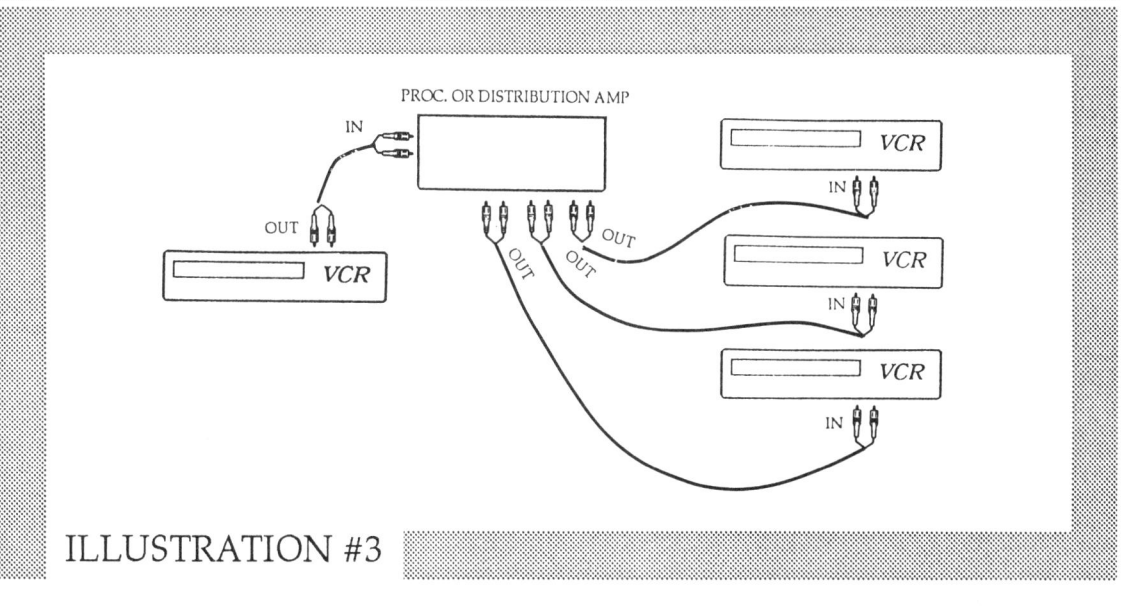

ILLUSTRATION #3

CAMCORDER ACCESSORIES
(The Extra Goodies Section)

VIDEO TRIPODS

Tripods are not created equal. Video tripods are different than camera tripods in several ways. Video tripods are taller than camera tripods (or should be). They should have a fluid head for smoother operation. You don't have to pan with a photo camera as you do with video. Try to get one with a quick release (this saves a lot of time). It should have a middle support brace. Always use a heavy-duty video tripod for VHS or S-VHS Camcorders. Take your video camera with you to the store to try it out. It will cost you about $49 to $99 for VHS-C or 8mm and $99 or more for VHS or S-VHS. No joke. There is a 99% chance that, if the tripod falls over and the Camcorder hits the ground, it is going to go to the video hospital. This is one thing you don't want to be cheap about.

CHARACTER GENERATORS

Most Camcorders are equipped to handle a character generator or video titler. On this, I say try before you buy. Some are good, some not so good. Most are equipped with 10 pages of titling, upper and lower case letters, numbers, and some with different colors. Note: almost all of them can only be used before filming or during filming, not days later to superimpose on the film. Another way to add titles is via computer. To superimpose titles onto pre-recorded video tape you must have a Gen-Lock on the computer, for example, while adding a person's name after filming. You can, however, add titles between scenes if you have video/audio insert editing capabilities on your Camcorder. *(See also the Direct-ED Plus Section of this book for another alternative for adding titles and graphics).*

EASY ANIMATION AND TITLES

So you want to save money on buying a character generator or titler, but want to have neat titles. Try using a titling video tape. That's a video tape with titles such as Happy Birthday, Merry Christmas, Chidren Growing Up and Our Vacation. One of the best I've seen is from Easy Animations, Inc. You get over 400 animations, graphics, titles, and sound effects for about $30.00. Not too bad to get you going, is it?

One of the available animations has a mouse walk on the screen and strike a match, setting off fireworks which shoot all over. The fireworks end up spelling "4th of July." These animated segments usually last about 12 seconds. Easy Animations offers several volumes of animations, such as weddings, male names, sliding titles, animated titles and custom titles.

For more information, see page 79 or contact:

> Easy Animations, Inc.
> 2121 Normal Park
> Huntsville, TX 77343
> 1-800-552-6402

GYRO VIDEO LENS

Unlike separate telephoto and wide-angle lenses, the gyro video lens, made exclusively by Optex, offers the option of converting from wide-angle to telephoto simply by reversing the lens without the need to remove it from the Camcorder. This not only saves time and energy, but because you don't remove the lens, there is less chance of dropping or losing it.

HEMISPHERIC LENS

This is a next-generation wide-angle lens, which offers a very wide, distorted view, creating a fishbowl effect. The view covers almost 180 degrees. This is good for great creative special effects.

TELEPHOTO & WIDE ANGLE LENSES

There are a wide variety of these. For two very good reasons, please take your Camcorder in and try it before you buy:

1) You may need adapters for the lenses to fit properly.

2) You may get a distorted picture around the edges (called the fish-eye effect).

THE "MUST HAVE" ACCESSORY

Haze Filter: Do not use your Camcorder without one. It screws on the front of your lens to protect it (from rocks, kids, etc.). If your video store does not have one to fit your Camcorder, a camera store will. No professional photographer is ever without one. They only cost about $3 - $7 and they are the cheapest insurance in the world to protect your investment.

THE EYEOPENER™

I should have listed this under a section called "Must Have Also Section." It is a small device that hooks to the eyepiece of your video camera or camcorder that will enable you to avoid squinting. And, it lets you use both eyes.

It takes about one minute to install. (It folds away after use). It also allows you to use your peripheral vision to view your surroundings. I have used this like everything else in this book, and I endorse this product as a must-have accessory.

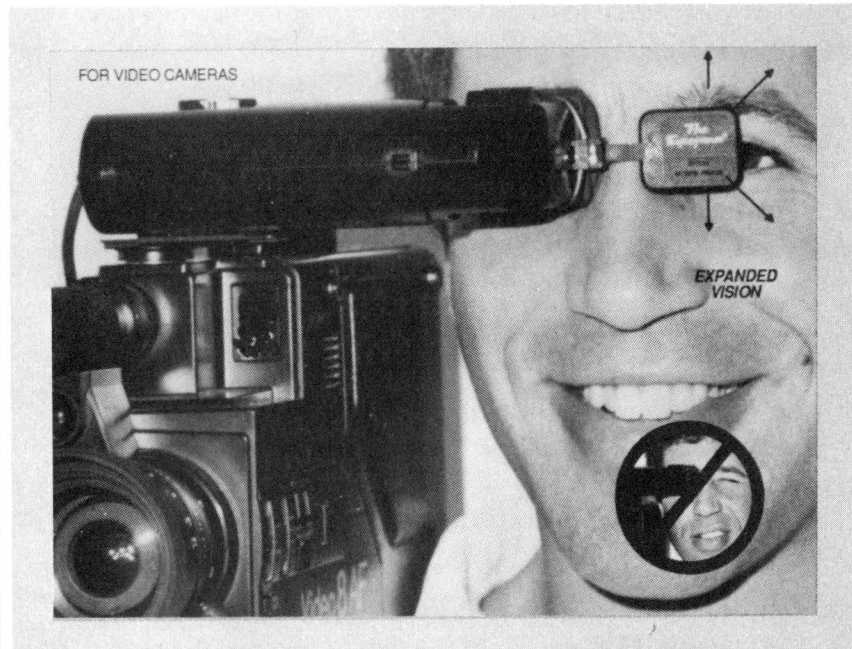

Don't Squint ... Relax And Use Both Eyes
FILM FOR HOURS WITHOUT EYE FATIGUE FROM SQUINTING
ADDS VISION AND AWARENESS WHILE FILMING

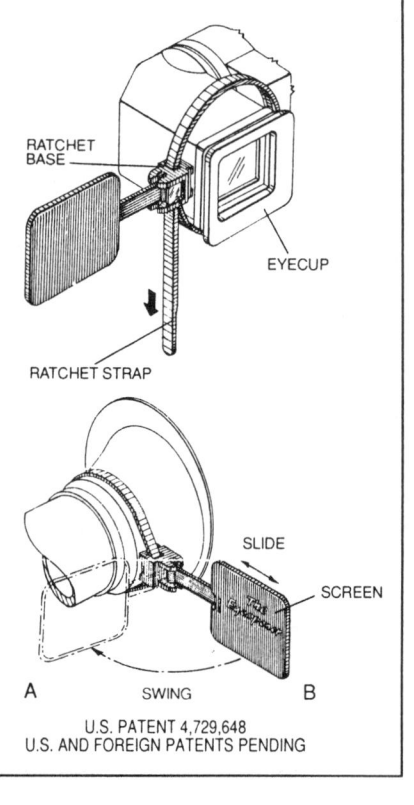

OPSIN
OPSIN INCORPORATED
18550 FIRLANDS WAY NORTH
SEATTLE WA 98133
(206) 542-7871

U.S. PATENT 4,729,648
U.S. AND FOREIGN PATENTS PENDING

EXTRA BATTERY PACKS (Are They Needed?)

That of course, depends on your use. If you just bought your Camcorder, don't buy any extra battery packs until you find out if you really need them. You might be able to use a car-cord adapter that would let you charge your battery from your car cigarette lighter (I've only seen these for 8mm, but they may sell them for VHS also). When I used mine at Disneyland, I filmed till lunch, recharged my battery during lunch, and then finished the day with just the one battery. Neat, huh?

Note: The batteries in a Camcorder are usually NI-CAD type which means that you have to charge them and discharge them a few times to bring them up to full capacity. The other alternative would be a battery belt or pack with a car-cord adapter. These offer several times the capability of the standard battery and (a good point here), if you sell your equipment, you can use the battery pack or belt for portable radios, TVs, CDs or any new video equipment or Camcorders. So, think before you buy.

SOMETHING NEW

It was bound to happen. Finally, a company is about to come out with "Ready-to-Shoot Camkits." These will be put out by Optex Corporation. These camkits will include things like a video tripod, haze filter, camcorder how-to book, special lenses, etc. which is a good deal because you save time and money all the way around. It saves you time shopping around for one thing here and another thing there. It saves you money because you are buying everything at a group price rather than paying for each piece separately.

These packages will be out in early fall of 1989, so contact your dealer to get the package or contact Optex Corporation to find the dealer closest to you:

Optex, Inc.
1150 David Bldg. "C"
Elgin, IL 60123
(722) 742-6300

or In Canada:
52 Lesmill Road
Don Mills, Ontario M3B 2T5
(416) 449-6470

EXTERNAL MICROPHONES (Say it Again, Sam)

Of all the Camcorders with built-in microphones, the 8mm records the best sound (without motor noise from the Camcorder). Unlike the older two-piece units of days past, the Camcorder (or more specifically the VHS, S-VHS, and even the Digital S-VHS Camcorders) record a faint motor noise, even when recording in a quiet room. This weird hum drives you crazy if you are going to film interviews, for example. You almost have to have an external microphone (most Camcorders have a jack for an external microphone). You can even buy wireless, remote microphones now.

Let's talk about wireless remote microphones. This is one of the products that you should definitely check out at the store or try at home **before** going out to do a wedding. Because it is wireless, you may pick up interference from such things as CB radios. This is definitely one of those products where you get what you pay for. A good one should be in the $200 range, although you may find one for less that will meet your needs.

Let's get into wired microphones. You may or may not need a powered microphone (depends on your Camcorder). You need to take it to a store and try it out. The ones that you buy as an accessory from the manufacturer of the Camcorder are usually expensive.

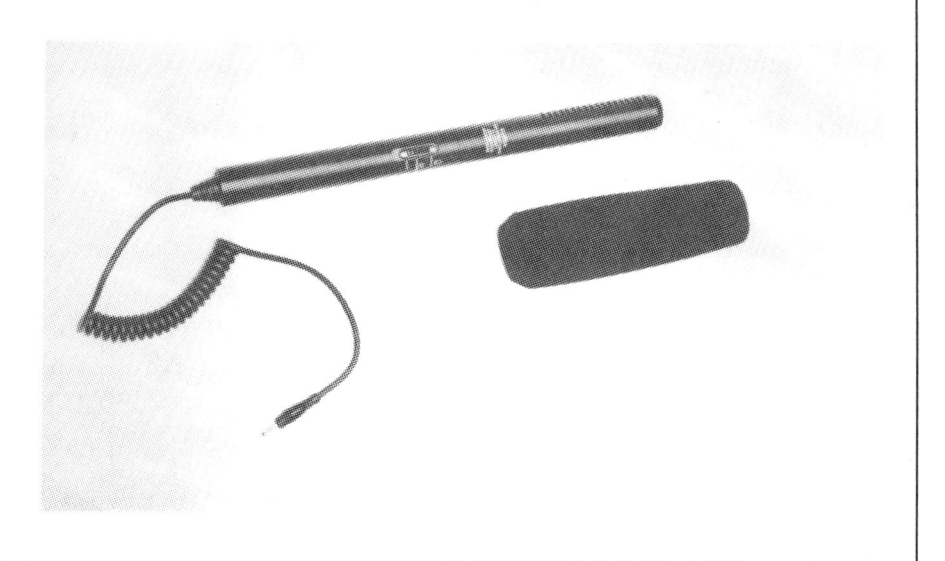

CARRYING CASES
(The Hard and the Soft of the Matter)

People have always asked me once they have purchased a video camera or Camcorder, do they need a case? If so, is it better to have a hard or a soft case?

The answer to the first question is a big YES. The question of which type is little more difficult. If you take a lot of trips (airplanes, boats, buses, etc.) get a hard case. It will protect your equipment better. But, if you putter around the house or take short day trips, you might want to consider a soft case (what I call a day bag). Now, the bag does not have to hold everything you own for the Camcorder (the hard case should). It needs to be able to carry the Video Camera, batteries and, most important, the battery charger. Make sure that the soft case is padded, especially on the bottom and outfacing side.

What the market needs (OK manufacturers, listen up) is what I saw at Disneyland in 1988. I met a man who had a digital S-VHS Camcorder with a cover fastened on by velcro made of the new breathable material that is used on deck model VCRs. You know, the material that says it has 5,000 breathing holes so that it will not overheat your VCR. I thought that it was the greatest thing since peanut butter and sliced bread. As I got down on one knee with my scratched and bruised S-VHS Camcorder, I asked him where he bought it. He put his hand on my head and told me that the stores do not sell them so he made it. Before I could hand him the contents of my wallet and my class ring, he sighed, turned, and dissappeared into the Submarine Explorer. This is a true story (except for the wallet and ring). If only the market had something like he had. It reflected heat, was able to breathe, was padded, and it was very easy to put on and take off. Now, that's the way to go.

VIDEO BATTERIES
(or Jingle Bells, Jingle Uh Oh!)

Many people buy their video cameras or Camcorders just before Christmas or other special occasions, not knowing or not being told of the Dragon. The Dragon is your battery pack. Battery packs almost never come charged enough for more than two or three minutes of use. Nobody tells you this or puts up a sign in the store warning you of the fact. So, if you are like 99% of the people that bought a video or Camcorder just before Christmas, you took it home, wrapped it up, and then on Christmas morning put it together, and probably filmed about two minutes before the Dragon got you. I would like to apologize for there being no warning sign in red letters on the outside of the box.

Warning: Before using this equipment you must charge the battery.

This section is not going to make me any new friends at the manufacturing plants, but I thought it had to be said.

P.S. It will take two or three chargings, draining down and recharging each time, before the batteries will be able to hold a full charge.

Speaking of batteries, you may want to buy extra batteries for your Camcorder. You can usually get them in two time lengths, one and two-hour. The one-hour battery does **not** mean that it will last a full hour if you turn the Camcorder on and off frequently and use the auto focus and zoom features a lot. It will probably last one-third to one-half less time than you expect unless you film straight through without using the power zoom or auto focus.

LUX: WHAT IS IT?
(One Candle Too Low)

It seems that one of the only things people know when looking for a Camcorder nowadays is to look for the LUX rating and get the lowest LUX. The misconception they have is that they think that with a seven LUX camera they can record a candle on a cake. At least it looks that way on TV. If you try this it will turn out very colorless, dark and very grainy looking. As an example, lets go down to one LUX. First, no one will confirm this one LUX rating. Even most of the owner's manuals say "approaching 1 LUX". In the owner's manuals it even suggests that optimum light intensity on the optical image is 1500 LUX which is 150 footcandles. This rating was taken directly from an owners manual of a Digital S-VHS current model Camcorder (Top Line).

I am not trying to project an air of gloom. I am just pointing out that we still cannot record properly in very low light conditions. If you plan on buying a Camcorder for a lot of low light conditions, plan also on buying video lights to go with it.

VIDEO LIGHTING

There are three types of video lighting: AC, DC, and AC/DC. The AC type usually ranges from 150 watts to 600 watts or more. These are primarily used in a studio situation.

The DC type usually ranges from 25 watts to 150 watts. The thing to remember here is that video lights **eat power**. I had a 150-watt light that ate up a six-hour battery in 50 minutes.

I suggest you stick with lights in the 35 to 100-watt range. Also, make sure your light has a guard over the bulb. There are two reasons for this; first, to prevent the light from damage and, second, (and by far the most important) if you touch the light bulb with your bare hands the oil from your fingers will cause it to either burn out immediately or within a minute. Replacement bulbs are not cheap.

The third type of light is AC/DC. This type of light will usually give you the option of about 25-50 watts on DC and from 50-100 watts or more on AC.

Since everyone's situation is different, it is hard for me to recommend one particular light that will fit every need. All I can say is research before you buy.

WARNING: Anything worth saying is worth repeating. **Do not** touch a video lightbulb with your bare hands.

TRACKING CONTROL
(Or Now You See It, Now You Don't Section)

Even though VCRs are set at the factory regarding the video tape speed, there may be slight differences from one VCR to another. The tracking control is used to adjust your VCR to match that of a previously recorded video tape from another VCR. For example: if there is sound but no picture or the picture is noisy or contains distorted streaks or horizontal lines, adjust the tracking control left or right until the problem goes away. If the problem persists examine the video tape for damage (see section on Examining a rental tape). If the tape is not damaged it is possible that your video heads are dirty or even worn.

NEWEST HI-TECH
(Tomorrow Here Today)

With the onslaught of people buying their own video cameras, and, after reading my book on how to make money in video, people will need new types of video equipment. The manufacturers and inventors have heard the cries of the consumer for new types of support equipment to help them on their quest. Let me point out three shining examples and where to locate them.

DirectED PLUS

This unit is a true breakthrough for the home enthusiasts of small-to-medium sized video production businesses. It is a computerized edit controller, special effects generator and a character generator all in one unit that will work with virtually all wireless remote control VCRs. It is extremely easy to use. Everything is step by step, on-screen. There are several types of fonts, and graphics to choose from with new graphics libraries and fonts that you can add to later. There's a help line and even a newsletter and user groups. You are in good hands for all the support you need.

To further give you an idea of what an incredible piece of equipment this is, let's talk money. I have personally spent over $12,000 and haven't got as much capability as this unit alone. For the people that want to see the specifications before they can believe; here they are:

The price is just as amazing as the product. The unit is currently priced at $549. That includes everything you need to get going, including your first graphics and font package. Additional graphics packages will be available for about $49.

Also, from Videonics is the ProED. It is like the DirectED, but is designed more for the professional, business, and advanced

Videonics *DirectED PLUS* video editing system

DirectED PLUS is a complete video editing system designed for anyone who owns a camcorder. It works with virtually all VCRs and camcorders and combines automated assembly editing with titles, graphics, and special effects, all in one compact package.

***DirectED PLUS* is a video editor!** Use the wireless remote control to mark your favorite scenes. You can mark hundreds of scenes on one original tape or on many. You can view each scene, fine-tuning the marked points, and rearrange the scenes. *DirectED PLUS* will remember the scenes in its permanent video library. It will automatically assemble them into a finished production, complete with titles, graphics, and special effects, at the touch of a button!

***DirectED PLUS* is a title generator!** Add brilliant titles to your videos using *DirectED PLUS*'s built-in title generator. Choose from 12 title styles and 64 colors. Each line can be a different style and color. You can superimpose titles on your video, on solid color backgrounds, or on a "graphic."

***DirectED PLUS* is a graphics and special effects generator!** Add your choice of 22 colorful, digital graphics, such as a birthday cake or wedding rings. Combine graphics and titles with any of 17 unique special effects. You can select the title and background colors independently, from a palette of 64 colors!

***DirectED PLUS* uses the equipment you already own!** All you need is a VCR with wireless remote control and a camcorder (or second VCR) capable of playing your original tapes, and a television/monitor with separate video and audio inputs. *DirectED PLUS* works with any video format — VHS, Beta, 8-mm, Super VHS, etc. The play and record units can be the same or different formats.

***DirectED PLUS* is automatic!** You create and edit video movies using the Videonics remote control, choosing actions from on-screen lists (called menus). A clear, easy-to-read manual guides you every step of the way. On-screen assistance is available at your fingertips with a press of the HELP key.

The production process is automatic, too — *DirectED PLUS* manages the details, telling you which original tape to insert and when. Even if a complicated production takes hours to complete, requiring that the original tape be rewound many times, all you need to do is follow the on-screen instructions. You don't need to wait around while the production is being assembled — a blinking light tells you when *DirectED PLUS* needs your help.

The quality is fabulous! Videos made with *DirectED PLUS* are always second-generation — copies of the *original* tapes, never copies of a copy. The graphics, titles, and special effects are digitally generated.

***DirectED PLUS* is surprisingly affordable!** For a fraction of the price of a typical camcorder, *DirectED PLUS* turns your home videos into polished, professional productions!

Prices and specifications subject to change without notice

hobbyist. ProED controls two VCRs simultaneously and offers computer-controlled assembly and insert editing.

The DirectED and the ProED are both totally controlled from a wireless remote control. But the ProED can also be controlled with a mouse or keyboard. It can control the editing from the keyboard as well as the remote control.

For more information and your local dealer please contact:

Videonics
1370 Dell Avenue
Campbell, CA 95008
or call (408) 866-8300
or 1-800-338-EDIT
FAX: (408) 866-4859

In Canada: Optex
52 Les Mill Road
Don Mills Ontario
M3B2T5
800-387-0275

HOW TO MAKE MONEY IN VIDEO
(Long Green Section)

With the quality of video cameras today, you can make a large amount of money with a Camcorder. I started my videotaping business with just $300. I placed some ads in the local newspaper saying that I did film-to-video transfers, transferred photos and slides onto video tape, and filmed houses for insurance purposes.

My first job was to do a wedding. I rented a video camera outfit from the local cable company for $45. I charged $350 to do the wedding, made two copies (one for $20 with a case and one for $15 without a case), spent $3 for gas and got free food. Tapes and cases cost me $20 and I walked out with $280 clear in my pocket. Not too bad for a day's work.

The next job I got was to transfer slides onto video for $640. I bought a new slide projector for $210, a special adapter for a video camera for $120, went to a store and put a video camera on my credit card (payments $32 per month), spent $20 for background music, spent $10 for tapes and cases. I set aside $96 for three month's payments on the new video equipment, worked a day and a half on the job, and walked out with $194, a slide projector (forever), slide adapter (forever), and three months payments on the video camera.

Within four months I opened my first store, and within two years my business was worth over $750,000. Of course, all of that was not just video production since we also sold video equipment.

It was not quite as easy as I have made it appear. At the time I was doing this, I worked full time also with the cable company. *All* the money from the video production went back into buying more equipment and advertising, etc. But, there is no reason why everyone of you can't make money in video, whether you want to make a full-time business out of it or just pay for your video equipment. The money is there if you know how to do it. I am going to unlock the secrets to home video production (the right way) so that you can give a quality product at a fair price. Please, for your credibility and for my peace of mind, follow these guides to the letter. Do not cut corners. Give a quality product.

STANDARD TAPE SET-UP
(Crawl Before You Walk)

Standard tape set-up is a must for any good production.

1) Put a blank video tape into your Camcorder or portable video deck unit.

2) Leaving your lens cap on, record about 20 seconds on the front of the video tape (if you can disconnect or unplug your microphone, do so). This method is used to eliminate the drop out at the beginning of the tape as on rental movies.

3) Put your name or company name and phone number on the tape using your character generator or computer. This is called a stripe or striping the tape.

4) Put your main title on. Use something like "Smith Family Album". Run that title for about 10 seconds.

5) Put the first subtitle on. Something like "Fun at the Beach". Put in as many titles as you need. Let the last title you put on run for at least 30 seconds before you begin the real filming.

6) If your client does not want music on the video tape (photo, slide, or 8mm transfers) make sure the microphone is disconnected or turned off during the transfer process.

This standard tape set-up should be done on all of your tapes whether you are doing weddings, births, wills, or whatever.

TRANSFERRING STILLS ONTO VIDEO TAPE

Equipment Needed
   1) A camcorder or two-piece video camera and recorder with video and audio insert editing, plus a macro setting. Preferably with a flying erase head.

   2) A good tripod.

   3) A stereo.

   4) Nice easy listening music.

   5) A character generator (one that either attaches to your video camera or by computer) or, as a last resort, hand-made titles.

Price to Charge
Here is a formula for a fair price to charge (you may deduct 20% off for family or friends, but for heavens sake don't do it free). Charge a one-time set-up fee for the first blank tape and the first 20 photos. Charge $4.50 for each additional 20 photos. Charge $15 an hour for adding music. Charge nothing for titles, if they keep it simple like a main title and four or five subtitles, or charge $1 per title until the character generator is paid for. The storage case is free. Buy a nice case (about $.50 to $1.00). Type in the title. Do not write the title and put it in the cheap sleeve that the video tape came in. It looks unprofessional and very trashy. Unless you have a video tape sample of your work, I would not ask for a deposit on the work, and would tell the customer that if they don't like it, they don't pay. Of course, you do not give them the video tape, and never let them keep it for a day to preview. Just play sections for them until they are satisfied that it is a good job.

How to Transfer Photos

Put your video camera on your tripod and hook it up to your TV set or monitor/receiver so you can monitor the transfer. This works a lot better than watching through the eyepiece of your video camera. Next, set the video camera by the edge of a table or desk (the edge of a table works best). Put your video tape in and put in your main title with your Character Generator. A title like "The Smith Family Album" usually works best. You may also want to include your name and phone number before the title for future orders from friends or relatives that may see the tape. Next, put in the first subtitle "1945 to 1947" or "Summer at the Beach", etc. Now, take the first photo and place it on the table.

Set your Video Camera vertically with the lens about an inch from the photo. Put your Video Camera in the Macro mode. This is done by putting the auto-focus on manual and then pushing or pulling on the macro button located on the lens housing (see your owner's manual for exact procedure). Manually turn it left or right until the picture is in focus. Next, raise or lower the Video Camera with your tripod until you get as much of the picture in as possible. At this point you may need to refocus. Also, you may need to add light from a table lamp or other source. Since VCRs record a square picture and photos are horizontal or vertical you will not be able to get the whole picture in unless you use a Special Effects Generator (about $450) or show the outer edges of the photo. But, you will get over 90% of the photo and most, if not all, of the action.

When this is done and it looks good and in focus on your TV or monitor/receiver you are ready to film. Play back your titles (remember your standard tape setup). When you get to the last title at the front of the tape, count 12 seconds and put the unit on pause. Set the unit on Record Pause. Take the unit off Record Pause by pressing the start button or Record Pause button located on the hand grip of your Video Camera or Camcorder. Count about 10 seconds for each photograph. With this time count you can put about 700 photos on one video tape. Always record in the SP or two hour mode. Put the unit back on Record

Pause after the 10 seconds and change photos. Now you are ready to go again.

Note: you cannot keep most Video Cameras in Record Pause mode for more than 4 minutes without the unit automatically shutting itself off. So you better hurry or you will have to set up the Record Pause process over again from where you left off.

Continue this process until all titles and photos have been put onto the video tape. Put a nice "The End" title on and then do an audio dub with music. (See how to add music section). Put it in it's nice case and you are done (except to collect your money).

How to Transfer Slides

Equipment needed:

The equipment needed here is much the same as for transferring photos, plus either a slide projector or a Video Camera slide adapter. Also, a large 11"x14" or 16"x20" piece of smooth, white paper or thicker stock.

Price to charge:

Same as for transferring photos.

How to transfer slides:

The process is almost the same as in transferring photos except a little easier. It is not absolutely necessary that you have a slide adaptor or what is called a Tele-Cina converter. It is your choice. If you do the transfer using a slide projector, set it up like you would for normal viewing. Attach the smooth, white paper to a wall with tape or tacks. Adjust the picture and focus to fill about 90% of the newly-formed screen. Set the Video Camera directly behind the slide projector and focus manually until you get most of the picture in. Take your stripe tape (see Standard Tape Set-up section) and follow the steps for recording photos (see How to Transfer Photos section). You may or may not, when using a revolving slide projector, want to let the Video Camera run continuously and show the momentary blank space after each slide, instead of using Record Pause after each slide. It depends on your personal preference.

HOW TO TRANSFER 8MM, SUPER 8MM, OR 16MM FILM ONTO VIDEO TAPE
(The Old to the New Section)

Equipment Needed:

The equipment needed is much the same as transferring slides, with the exception that you will need an 8mm, Super 8mm or 16mm projector with an adjustable speed control. This is very important.

Price to Charge for 8mm or Super 8mm:

The price should be as follows: $25 set-up fee (includes first blank tape plus 200 feet of film). Then charge $19 for each additional 400 feet. You may break the pricing down into smaller increments if you wish. For instance, if your client has 600 additional feet, charge him $28.50 extra, etc. Here's the catch with 8mm and Super 8mm. If you look at the film box, it may say something like 25 feet. This is because you filmed on one side, turned it over and filmed on the other side. However, when it is processed, the developer will cut the film and splice it together to make 50 feet, so you should charge your client for 50 feet.

Price To Charge For 16mm Film:

$35 set-up fee. This of course includes the first blank tape and the first 200 feet of film. Then charge $24.50 for each additional 400 feet. This is due primarily to the fact that a 16mm projector will cost more than an 8mm or Super 8mm projector.

HOW TO TRANSFER MOVIE FILM ONTO VIDEO

Set up your equipment the same as for slides with the Video Camera behind the projector. Set the projector on a smaller object like a coffee table, as a film projector is taller than a slide projector. The secret to a good film-to-video transfer is speed. If you do not adjust the speed of the moving film you will probably get a flicker or what looks like light bars moving progressively up the TV when you play back the video tape. Film and video run at two separate speeds so you have to speed up the film to get rid of the light bars or flicker. Don't worry about any adverse effects created in the viewing by the speeded up film. Grandma will not be running 90 miles an hour. You won't even notice it. Follow the rest of the procedures for filming that are in the transferring photos section. Always hook up your Video Camera to a TV or monitor/receiver to monitor what you're doing. Once again, never attempt to make a transfer of nitrate based 16mm film. Smell the film if you are not sure, and look for a pungent odor. If you are in doubt, take it to a professional.

HOW TO VIDEO TAPE A HOUSE FOR INSURANCE PURPOSES
(or Big Bucks, Knock on Wood Section)

Equipment Needed:
Video camera or Camcorder with macro settings.

Price to Charge:
For an average-size, three-bedroom house, charge $75 if you can do it in about an hour, $20 per hour for each additional hour.

Step-by-Step:
First stripe the video tape as usual (see standard set-up), then put down the date that you did the recording on the front of the tape. Use the full date, month, day and year. Put this also on the back of the tape when done.

When you video tape the house, start in one corner of each room and pan the entire room slowly, including drapes, wallpaper, etc. Do not forget closets or bookshelfs (get the titles of books on video tape). Film light fixtures, towels, everything. Ask if there have been any improvements made on the house. A good example would be additions to a small bathroom; a wooden towel bar ($35), a wood or brass medicine cabinet ($100 to $250) or even a wooden toilet seat ($50). Don't laugh, it all adds up.

Do the garage and/or storage shed. Tools are not cheap. When you're done, make sure your client takes the video tape to either another relative or to a safety deposit box for safekeeping. Believe me, if something happens, like a fire, they will get a lot more money from the insurance company than if all they had were a couple of photos of their TV or stereo.

If items have serial numbers, use your macro setting to record them. Also, don't forget jewelry or antiques. Ask how much it is worth, not how much they paid for it.

You can ask the owner to talk into the microphone about these things as you go along.

HOW TO MAKE A VIDEO WILL
(Dead But Not Forgotten)

Equipment Needed:
A video camera or camcorder equipped with a date and time function (this is a legal must) and a tripod.

Price to Charge:
$75 if the will takes an hour or less. $20 per hour after the first hour.

Step-by-Step:
There are two reasons to make a video will.

1) It is a legal means to handle your clients estate. You must however encode a date and time on your video tape to make it valid in all 50 states.

2) It gives your client the ability to talk to his/her family and friends that he/she leaves behind.

I have done several video wills with people that have children. These are generally set up in sections. For example:

1) Reading of the will.

2) When the children are 12 years old.

3) When the children are 18 years old.

4) When the children are 21 years old.

They include their hopes and dreams for their children, on a one-on-one basis. This may seem a freak of modern technology, but just think about it a minute. If given the opportunity to die suddenly and miss a final farewell, or to die and be able to spend a day or two to have these talks, which would you choose?

You start the video will the same way as your standard tape set-up. You may also want to include on the front of the video who is to watch the video. Then, with the year, date and time being encoded on the video tape, have the person introduce themselves with full name, current address, and social security number. Next, have them dispel all other video or written wills that now exist. At no time from start to finish should you stop the video camera or Camcorder. You may *not* edit out mistakes or anything else. This would invalidate it as a legal document. Keep this in mind before you start. You may want your client to make a list of things to talk about before starting. At the end of the video will, have your client state that it is the end and nothing else should be added. Make sure that you put the video tape in a case and have your client put it immediately in either a safe deposit box or turn it over to his lawyer or executor of his or her estate.

INTERVIEWING FOR MONEY
(One Video is Worth 10,000 Photos)

Equipment Needed:
Video camera, tripod and, if you are doing the interview yourself, a remote pause button for the Camcorder.

Price to Charge:
A good fee is $65.00 for the first hour (assuming it will take you 1/2 hour to set-up and prepare). Then charge $25.00 for each additional 1/2 hour.

Doing the Interview:
First, do your normal videotape set-up. Second, if you have a blinking red tally light on the front of the Camcorder, COVER IT UP! It is very distracting to the interviewee. Next, build into the interview today's date and what date you are talking about. Just mention the dates as you progress in time. Most of the time you will be interviewing older people so it is good to explain what you are doing so they will be at ease. (Don't assume they know about video.) Try starting with what they remember about their childhood, for instance, did they buy or bake their bread? Did they have electricity or an outhouse? (Some of the funniest stories I've heard are about outhouses or how they got water into their house.)

Next, ask them about their grandparents' accomplishments or what country they were from, etc. Some more examples are:

- Did you have a good childhood?

- What kind of games did you play?

- What did you do at parties?

- What did you want to be when you grew up?

- Did you achieve what you planned?

- What would you hope for your children or grandchildren?

- Etc.

These are just some guidelines. You may also want to get a professional guide specific to this subject. The best I've seen is put out by:

        Living Family Albums
        25935 Detroit Rd.
        Westlake, Ohio 44145
        (216) 892-6826

HOW TO VIDEOTAPE A WEDDING
(With This Video Section)

Equipment Needed:
Video camera or Camcorder with audio dubbing capability, tripod, wedding music.

Price to Charge:
$365, Including transportation to wedding and reception (limited to 50 miles round trip), and up to four hours of filming (wedding and reception combined). $25 per hour for each additional hour over four.

Step-by-Step:
There is nothing better than a wedding on video tape. You get all the movement, vows being said, and sounds that can never be captured on still photos. But, of all of the wedding videos that I've seen, over 95% were so bad that the family member, friend, or so called professional should be sued or shot. This is one of the types of filming that cannot be redone. You only get one try at it. So, do not attempt this type of video filming unless you feel confident and competent with your equipment. Remember "if you can't stand the heat, stay out from behind the camera". For all you people out there that would consider renting a video camera and letting your Uncle Fred or neighbor Bill film your wedding, let me ask you this, would you let your Uncle Fred or neighbor Bill take still photos of your wedding? Enough said.

If you're going to film a wedding, do it right. First, ask the bride and groom for about 10 or 12 photos each from the time they were born till the present. Then get about four photos of them together before the present. Put these on the front of the video tape after the main title. Remember to use the maiden name of the bride to be until after the wedding when her name will change to her married name.

On the day of the wedding, take an outside shot and some inside shots of the church with your video camera or Camcorder. Then, do a short interview with the bride and groom. Also, at this time remind them to speak up. It has been my experience that in almost every wedding the bride and groom will say the first two words at a normal volume and then almost whisper everything else. Also, at this time make sure you tell the Maid of Honor, Best Man, and Bridesmaids not to block the video camera. The best angle to film from is to be on the left-hand side as you face the alter. Do not film from any sacred areas. Check with the church before you set up. Normally you do not need to film everyone being ushered in. This takes up an enormous amount of video tape each time. You will have plenty of time to film who is there during the ceremony. Here are a list of ethical Do's and Don'ts when filming a wedding.

1) Do not move your tripod around. This will distract and interfere with the wedding. Use your zoom instead.

2) Do not use lights. Have them (if necessary) turn up the lights in the church. This is a very big distraction.

3) When filming the wedding make sure that you zoom in on the ring ceremony.

After the entire ceremony has been completed it is not necessary to film everyone leaving. Just the bride and groom and maybe the parents as you fade out the scene.

Next, is a neat section which I call video-togrophy. When the regular photographer is taking his or her photos of the bride and groom you can be filming little shots between them. You must be quick to get these shots. Always make sure that the still photographer is not in the scene and that you do not get in his way. If you couldn't get a good shot of the ring ceremony, now is your chance. When this is over you may want to put in a title or leave a space to put one in later (you can only do this if you have a video camera with video and audio insert editing). Use something like "Congratulations, Mr. & Mrs. Smith. Now, go on to the reception.

At the reception, assuming that you drove fast to beat the mob before they converged on the food, film the wonderful food and then the cake. Be creative. Then, film the crowd. Make sure that they are not eating or drinking. There is no better way to ruin your video than showing Uncle Fred with a can of beer in his hands. Next, film the people (section by section) around the tables. This may take two or three set-ups to get everyone in. Again make sure that when you film them, they are not eating or drinking.

Mill around a while, but don't film all the time. About five minutes is plenty. Next, film the following:

1) Toasts

2) Cake cutting

3) First dance

4) Join in dance

5) Throwing of the bouquet

6) Taking off and throwing of the garter

7) Putting the garter on the one who caught the bouquet, done by the person who caught the garter (if applicable).

Then as a good closing film the presents and the bride and groom waving.

As a professional, your work begins here. Here's where we do the audio dubs and where to add them.

1) Dub music from the main title through the still pictures of the bride and groom to be, through the shots of the outside

and inside of the church, right up to the interviews. Refer to your camera's audio dubbing section for your video camera's particular audio dubbing instructions or check the section on Adding Music in this book.

2) Dub in more music after the ceremony, during the videography session, through the filming of the food, cake and people, right up to the first Toast. This is done because the voices in these sections will be quite noisy and hard to understand anyway.

3) Dub in more music after the putting on of the garter, through to the closing titles.

Pick music for your dubbing that is either of the wedding type or soft selections picked by the bride and groom. If they ordered extra copies make them now.

I used to finish the wedding, go back to my studio, finish the audio dub and be ready to show the finished product by the next morning. By doing this, the sales of copies more than tripled, usually because the out-of-town guests were still in town to see it. Also it brought all kinds of other business because it looked very good and was done in record time. Good luck!

IDENT-A-KID
(Every Six Months)

Equipment Needed:
Video camera or Camcorder (hopefully with date encoding capability) with macro and a tripod.

Price to Charge:
If you have your own store front or office where the client can come to you, I do not feel that you should charge anything if your client supplies you with the video tape. Or, just the cost of the video tape, if he doesn't. If you do not have your own store front or office and you have to go to your client, you should charge a small amount for gas and travel time. The reason I suggested to not charge for this service is because our youth of today is our future of tomorrow and should be protected at any cost. You may even want to advertise this in a local newspaper.

Step-by-Step:
Set-up a video tape as usual. Then, put in the child's name, date of birth, the date you are filming, color of eyes, hair, and blood type. Have a measuring stick attached to the wall and have the child stand by it. Do a full shot, then zoom up to the face. Have the child smile. Have the child turn sideways, then smile. Have the child show his or her other side then smile. Ask the mother or father if the child has any unusual markings or scars. Film them close up with the macro setting on your camera. If the child has had his/her ears pierced, film them with and without an earring in. This is for later identification purposes. Have the child speak. This is for possible voice print identification purposes. The macro setting on your camera can also be used for fingerprint identification. Just have the child put his/her hand on an inkpad and then film using the macro focus of your video camera or Camcorder.

This procedure should be done every six months at least until the age of 18, and then your client should have it done once a year.

TRANSFERRING IMPORTANT DOCUMENTS

Equipment Needed:
Same as for transferring photos

Price to Charge:
The price to charge for transferring important documents is much the same as with photos with the exception that you should charge by the shot rather than the document. For example, if you are video taping a long legal document you may need to start and stop the Video Camera three times or more to get all of the document in piece by piece. So, charge by the shot.

Transferring Documents:
The procedure is virtually the same as with transferring photos. Make sure when you do this that you get all of the important papers such as birth certificates, stocks, bonds, etc. Also, make sure that your client puts the finished video tape in a safety deposit box or in another place other than where the important documents are. You may also want to put a date and/or time code on the tape.

BUSINESS VIDEOS
(Dawn of a New Age)

If You Use And Send Video Tapes

Gone are the days where a black-and-white Xerox copy of a flyer will get lots of business for your company. Gone are the days of your color brochure ending up in the circular file (trash can) before it got anywhere near the intended readers. Gone are the days of spending thousands of dollars to redo catalogs just to add a few new products to them and then having them end up where the color brochure did. Here are just some of the uses of video today.

It's hard to be in two places at once. Many businesses record their important meetings and distribute them to the other branches. This is a cost-effective way to have their new policies heard by all of their employees.

For demonstration of new techniques, training, or procedures, nothing beats video. It's a sales tool that is light years ahead of the competition.

Instead of sending out a catalog, send out a video of your products. To customers it is like seeing the product first hand. I have never heard of a company throwing a video into the trash can without seeing it first. People are curious.

Video tapes are important. They show that your company is not in the stone age, that your company is moving forward, not backward. That's where the people you want to do business with want you to be.

HOW TO VIDEOTAPE BIRTHS

Equipment Needed:
A video camera or Camcorder (it is helpful to have one with date and time function) and a tripod.

Price to Charge:
Usually $100 to $150 if the event does not last over three hours. Also, depending on what time of day or night. If it's before 5am you know to use the $150 charge. Work this out in advance with your client, so that he/she knows that an early morning birth will cost more and he/she isn't surprised with the extra charge.

Step-by-Step:
Do your standard tape set-up in advance, since you are on call and may not have time later. Interview your clients and put them at ease. If you have a woman on staff you may want to use her. Try to use the lighting that is available and not use extra spotlights if at all possible. Always act and conduct yourself as a professional. Do not make any copies unless requested to do so by the new parents themselves. You can (and should be) sued for showing any part of a birth to anyone else. Here are some pointers:

1) Try to film the monitor which displays the heartbeat of the mother and baby.

2) Stay out of the doctor's way. Use your zoom.

3) You may want to use manual focus as the doctor will be between you and the baby. This will cut down on the times that the video camera has to switch its auto zoom from one depth of field to another.

After the filming is done you may want to suggest that your client keep the video tape in a safety deposit box until the child is eighteen and decides for him or herself to see it.

ADDING MUSIC TO VIDEOS
(How Sweet It Is Section)

There are two ways to add music to the videos that you recorded with your video camera or Camcorder.

The first one (see illustration #1) is in the event that you do not have a Camcorder with video and audio insert editing. Unfortunately this can only add music to a copy since, on the original, you cannot change the audio without changing the video also.

Put the original video tape into the playback unit and rewind or fast forward until you get to the place on the tape where you want to start dubbing in music. Put a blank tape into the recording VCR and put the music that you want to add into the stereo (cassette, record player etc.) Go to the beginning of the music and put the stereo on pause. Go to the recording VCR and put

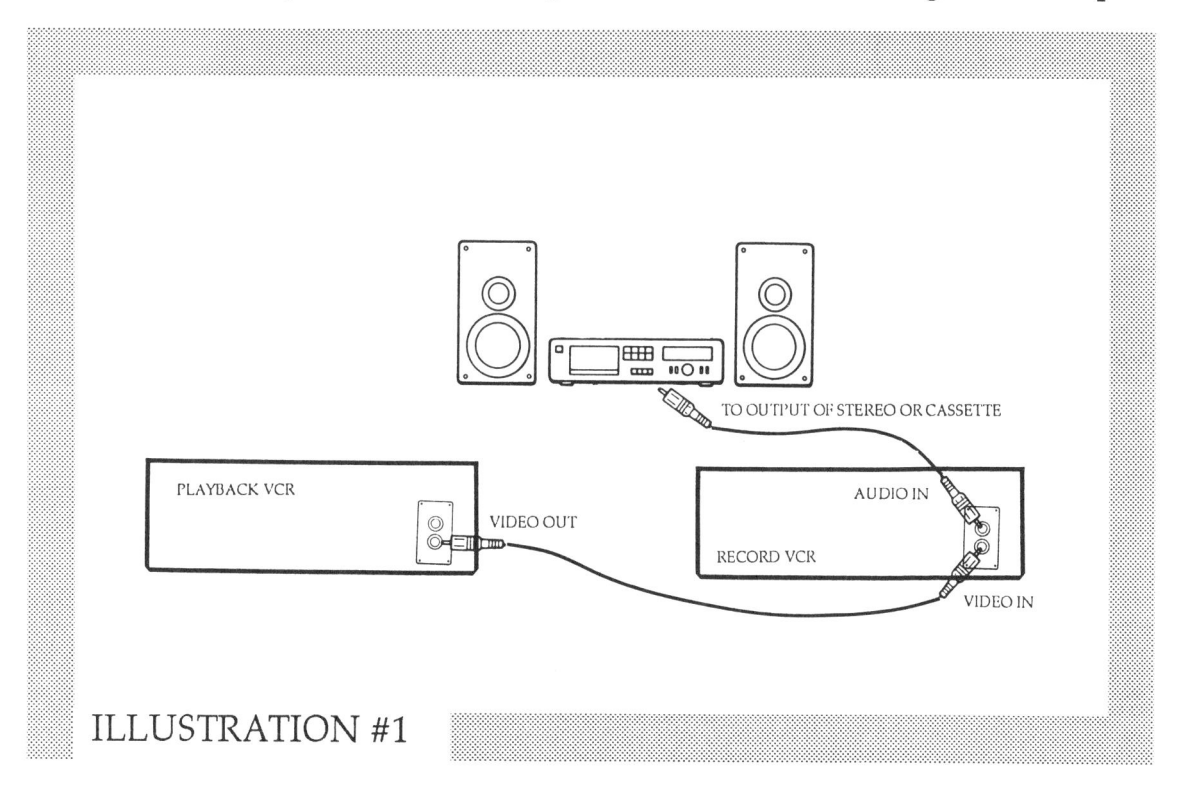

ILLUSTRATION #1

the unit in Pause Record Mode. Start the playback VCR and push the release pause of both the recording VCR and the stereo. Now you will be recording the video from the playback VCR and the audio from the stereo onto the recording VCR.

The second way to add music is by far the easiest and also the best because it can be done on the original instead of a copy. But you must have a VCR or video camera or Camcorder with an audio dubbing capability.

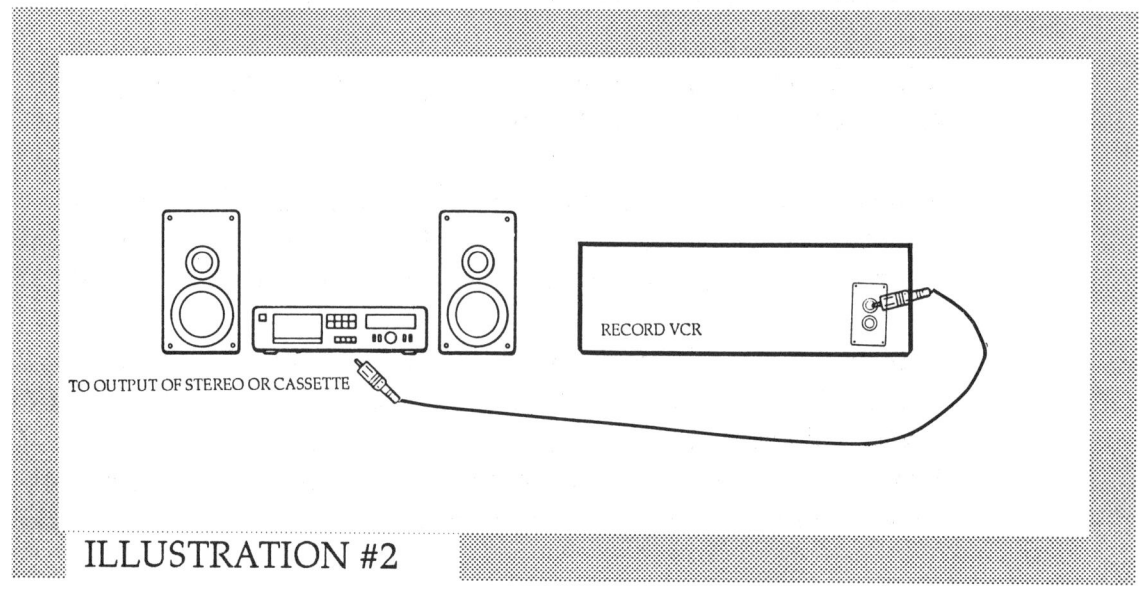

ILLUSTRATION #2

First, put the video tape (make sure the recording tab has not been removed) into the Camcorder or video deck with audio dubbing capabilities. Then, put the music you want to record into the audio source. Then, play the video tape until you reach the part where you want to add music. Put the unit on pause, press the dub button at the same time as the play button. This will put the music you want to dub in your audio source and put the unit on pause. Release the two pauses at the same time. Press stop on both units when dubbing is done.

HOW TO USE THE ZOOM
(The Closer You Get Section)
or (Let's Play Leap-frog)

With the advent of modern technology all video cameras or camcorders except the cheapest of the lot, (I know I'm not supposed to say cheapest, but least expensive. But, in video, I think cheapest applies well) have power zoom, usually six to one, eight to one, or ten to one.

You should not use the zoom for more than half of its capability without it being on a tripod or shoulder brace. The more you use the zoom the slower you should pan or move the camera also.

Especially, do not walk with the camcorder. There are two reasons for this. First, if you walk, the natural movement of your body will give you an up and down movement which will transfer to the tape you are making. Second, you won't be able to see where you are putting your feet.

I suggest that you use what I call the leap-frog tactic. First, place your feet straight forward about 12 to 14 inches apart. Then zoom out on your subject. If you can't get close enough, stop the camera, walk closer, and then place your feet again and start your zoom over. This will give you a better picture than walking up to your subject with the zoom out.

HOW TO USE THE MACRO LENS
(Incredible Closeness)

I love using the macro. Especially when I sell camcorders. I will take a dollar bill and put it right on the lens while using the macro lens and show people both sides of the dollar bill at the same time in perfect focus.

Focusing with the macro lens is done, manually even if you have auto-focus on your camcorder. The macro setting is located on the right side of the focusing ring as you face the camcorder. To operate it, push or pull the release mechanism (button) and slowly move it toward the bottom of the camcorder while looking through the eyepiece. You can get as close as 3/16th of an inch. The macro is used to transfer photos onto video tape or to shoot any small object such as flowers, butterflies, stamps, coins, etc.

HANDS FREE CAMERA SUPPORT
(Look, Ma, No Hands Section)

When you have to be on the move (on a boat, in a parade, etc.) and a tripod is not easily used, you may want a video camera support.

These come in several different styles and types. From the simple ones that just pad the Camcorder on the shoulder to the chest-pod type. The ones that just pad the shoulder do nothing to relieve the strain on the right arm which is gripping the camera.

So, my recommendation would be the chest-pod type. Since most people are buying full-sized VHS Camcorders, make sure the one you pick will support the weight of your particular Camcorder unit. As usual, try before you buy. Bring your Camcorder into the store to check it out first.

DEW INDICATOR
(From the Frying Pan Into the Fire Section)

The Dew indicator is a device built into deck units or camcorders to protect the equipment from excessive moisture buildup on the video heads. Going from an air-conditioned car to the very hot outdoors, or going from the very cold outside to a warm room, or putting a cold video tape into a VCR or camcorder, will create dew or moisture.

The Dew indicator will automatically shut the unit down, usually from about a half hour to an hour. Do not attempt to operate the unit until the Dew light goes out.

Always try to acclimatize the deck, camcorder or video tape to the new environment before use. This will avoid a lot of disappointment.

TIPS ON FILMING
(Icing on the Cake Section)

1) Try to film with the sun or light source behind you. Otherwise the subject you are filming will look darker than the background.

2) Try to use a tripod whenever possible. Especially when using the zoom.

3) Make sure your equipment and video tapes have a chance to warm up or cool down to meet conditions of the new environment before use. (See Dew Indicator Section).

4) Never walk with a video camera (See Zoom Section).

5) Make sure you carry twice the batteries that you think you will need. Especially if it is a hot day or if you are going to turn the equipment on and off a lot.

6) Never point the video equipment towards the sun even when the camcorder is off.

7) Always use a haze filter (See the "Must Have" Accessory Section). Also put the lens cap on the camera whenever it is not in use.

8) Avoid using colors like red or blue as backgrounds or clothes. They tend not to come out as well as you would want. Also avoid small pin stripes as they have a tendency to give a wavy effect.

SPECIAL EFFECTS
(Beyond reality)

If you have done any type of special effects with a still camera, almost all of them can be applied to your video camera.

1) If you put vasoline around the edges of the lens it will create a dreamlike effect. Don't put this directly on your lens, but put it on your haze filter. Of course, the vasoline can be taken off with very soft tissues or a lense cleaning kit. Don't submerge your whole camera to clean it. This could result in damage to your camera. (No joke. You have to tell some people about this.)

2) Using dry ice gives a fog-like effect. You would be surprised at the incredible effect that five dollars of dry ice at a wedding will create.

3) If you are filming a small object on the ground, try getting flush with the ground to make the object appear much larger than normal.

4) Another neat trick is to buy a Hex or Star filter. This attaches to the front of your camera and creates an effect that turns all the points of light that enter your camera into stars. So if you have three glints of light coming off of someone's hair, each of them will turn into a six-pointed star of light.

5) Still, by far, the best special effects are done with digital right in your camera. You need to buy either a digital camcorder or add digital from a digital VCR or special effects generator in order to take advantage of these.

VIDEO NEWSHOUND
(or How to Sniff Out Money Section)

At first, you may think that you can't videotape anything with good enough quality to be put on TV. Think again. At least twice a week on the news I see videos described by the newsman as "taken by an amateur videographer." Example: Recently, I saw film of a helicopter putting an air conditioning unit on top of a building. During this operation, the tail rotor hit something and the helicopter dropped its load and crashed. It was all captured by someone just like you.

Your local news station is now equipped to handle your VHS tapes. So call up your news station or cable company if you think you have something newsworthy or when you become good with your video camera and ask to cover small assignments that they might not have the time to cover. This is not how Barbara Walters got started, but I don't think the TV station or cable company will laugh at you if you approach them with the idea these days.

I have made TV commercials which were better than those put out by our cable company. I used my digital S-VHS Camcorder. (I have the tapes to prove it.) The picture was less grainy and my titles were better. We're not talking what I filmed or how I filmed it. We're talking the quality of what I filmed.

It's just a matter of being in the right place at the right time. It's also fun and you might make a few bucks, too.

GLOSSARY OF VIDEO TERMS
(Lookout Webster)

8MM - This refers to the tape or film width. There are now two types of formats with 8mm. They should have been called 8mm Video, 8mmV or something like that. To date, about 20% of the people in the USA think that 8mm Video has to be developed, because of the confusion between 8mm movie film and 8mm video tape. (No joke.)

AUDIO DUBBING - With the Audio Dubbing feature you can change the audio on an original video tape without changing the video portion. This is great at a party when you start and stop the camera often. This gives you the ability to go back and add some music to the whole tape.

CAMCORDER - This is a video camera and video recorder (VCR) in one unit.

CCD - This stands for Charged Coupled Device. It is an image sensor that replaces the old video tube. It is more durable and vibration resistant and does not need any warm-up time.

CHARACTER GENERATOR - This lets you add titles either during filming or at a later date (depending on what type of Character Generator you have). This term is also referred to as a CG.

DEW INDICATOR - This indicator will automatically operate if there is excessive moisture in the unit and will shut the unit down for a period of time.

EVF - This stands for Electronic View Finder. Simply put, it is the little TV in the eyepiece of your camcorder.

FLYING ERASE HEAD - Simply put, this gives you clean edits and cuts between scenes. It's a shame if you don't have it on your camcorder.

GAIN UP - You use this button in extremely low light conditions. The picture may become grainier.

HIGH SPEED SHUTTER - You can record fast-paced movement with the High Speed Shutter switch on. Then, while in the playback mode or slow mode, a VHS (VCR) will show you the fine points of any motion at 1/1000th, 1/500th or 1/250th of a second with no blurred screen image. I think this feature is more hype than anything. Also, it really jumps your LUX dramatically, so only use this outside with plenty of available light.

HQ - This simply stands for High Quality. It provides much better pictures than a non-HQ unit. Just about everything in VHS is HQ.

INDEX/ADDRESS SEARCH SYSTEM - With this system the unit automatically records an index and address signal onto the tape (invisible to us). You can also manually record an index mark later on a unit with this system. You can locate specific points on the video tape in the fast forward or rewind mode.

LUX - This a measure of brightness. One LUX equals one canlepower.

MACRO - This feature lets you view small objects incredibly close.

RECORD REVIEW - While in record/pause mode you can use the EVF (Electronic View Finder) and the Record Review buttons to check the last few seconds of video tape just recorded. This lets you make sure that your shot came out the way you wanted.

SPECIAL EFFECTS GENERATOR - This lets you add a multitude of special effects like color borders, strobe effects, screen dividing, etc. depending on the unit. This term is also referred to as an SEG.

TRACKING CONTROL - Even though the tape speed of VCRs is set at the factory, there may be a slight difference from one VCR to another. The tracking control is used to adjust your VCR to compensate for any difference in speed and match that of a previously recorded video tape from another VCR.

VCR - This term stands for Video Cassette Recorder. The video tape is in a cassette format.

VTR - This term stands for Video Tape Recorder. The video tape is in a reel format.

VIDEO DUBBING - The Video Dubbing feature added with the Advanced Rotary Flying Erase Head technology allows you to very cleanly replace a segment of video tape with a new segment without any lines of color bar distortion moving down the picture. This feature and the Audio Dubbing feature should be a couple of the top things you consider and look for before you buy your camcorder.

ZOOM - This feature makes objects that are far away appear closer.

Geneva Presents
Great Performers For Your Great Performances

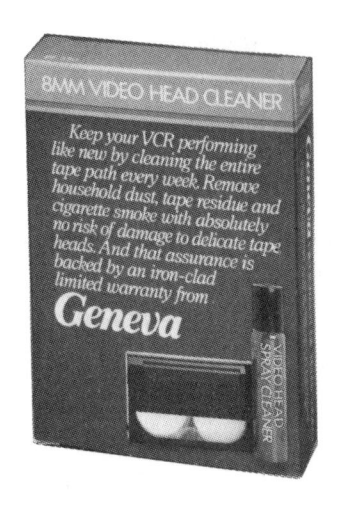

Capturing your family's Great Performances is why you purchased a camcorder. Now, make sure those special performances, your child's first steps, high school graduation, or your grandchild's first birthday party, are captured with all the clarity your camcorder can deliver by using the Great Performers from Geneva.

We know your camcorder will be used at home, at the beach or on vacation. Spending a few moments after each use to care for your camcorder will ensure it's ready to capture those Great Performances when they happen. It's easy with Geneva Head Cleaners and Geneva's Super Blast™ Air Duster to keep your camcorder performing like new.

Regular use of a Geneva Camcorder Head Cleaner helps prevent tape oxide build-up and safely removes other airborne contaminants that can have an adverse affect on the performance of your camcorder.

We designed these products to be compatible with the special technical needs of camcorders. With Geneva's exclusive performance guarantee, you can be sure that your camcorder will perform when asked to capture your Great Performances.

Call us at 1-800-328-6795, extension 382, or write us for the name of your nearest dealer.

"We Guarantee It!"

Geneva™

©1989 Geneva Group of Companies, Inc.
Geneva is a trademark of the Geneva Group of Companies, Inc., Plymouth, MN 55441
Super Blast™ is a Registered Trademark of the Geneva Group of Companies, Inc., Plymouth, MN 55441
In Canada distributed by Optex Corporation, Toronto (416)449-6470.

New!
The Grandparent's Video Interview Kit™

Interview parents and grandparents on videotape.

You get everything you need:
- The complete 92-question *Interviewer's Script*
- Illustrated *Camera Person's Shooting Directions*
- *How to Produce the Interview*
- *The Camera Person's Guide*
- *The Interviewer's Guidebook*
- Custom cassette labels
- And more.

It's fun to do, and a priceless gift for future generations.

This kit makes a great present for a special occasion, too.

Whether you own, rent, or borrow the video camera, **The Grandparent's Video Interview Kit** enables you to produce wonderful **videotaped interviews** of parents and grandparents. In just 60 to 90 minutes, you can capture their family histories, their life stories, their goals for the future--even their advice and wishes for their great-grandchildren. It's easy to do, and you'll make **video keepsakes** your family will treasure forever.

Do it for the grandchildren.

- **PHONE ORDERS:** Call **1-800-257-2002**, ext. **426**, Monday through Friday, between 8:00 am and 4:30 pm (Eastern Time Zone). Please have your VISA or MasterCard number and expiration date ready.

We ship orders via UPS within 72 hours of receipt.

- **(Quantity)** _____ Grandparent's Video Interview Kit(s) @ $34.95 each. $ _____
  Video One ($29.95, plus $3.00 Shipping & Handling, plus $2.00 sales tax.)

- **MAIL ORDERS:** Make your **check** payable, and mail this coupon to:
  Living Family Albums
  25935 Detroit Road, Suite 294,
  Westlake, Ohio 44145

- **Or use your** ☐VISA or ☐MasterCard (please check one).
  Card number _____
  Expiration date _____
  Name on card _____
  Signature _____

Money-back guarantee.
If not completely satisfied, return the kit within 30 days for a full refund.

- **Ship my kit(s) to** (please print clearly):
  Name _____
  Address _____
  City _____
  State _____ Zip _____ Phone # _____

LIVING FAMILY ALBUMS™

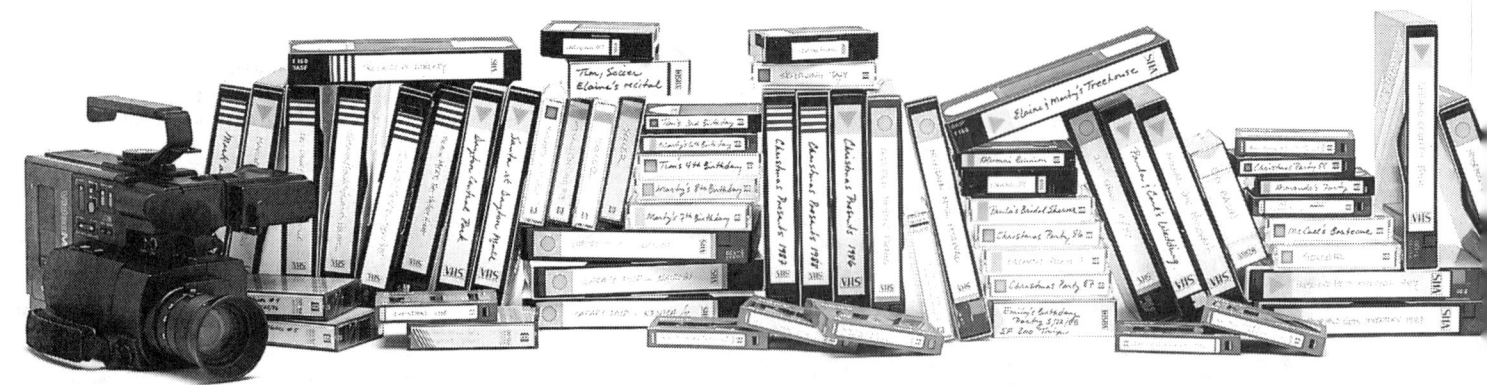

Your memories are worth keeping.

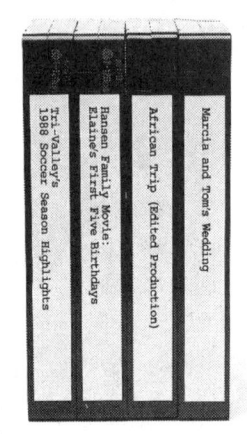

Make them worth watching with DirectED PLUS.™

Your home videos contain some great moments. And some not-so-great moments. Now you can make your home videos something you'll be proud to show with DirectED PLUS, the personal movie maker from Videonics.™

DirectED PLUS is a complete video editor! Now you can edit unwanted footage from your home videos, just like professionals do! Mark hundreds of scenes, on any tape, or tapes, in your home video library. DirectED PLUS will automatically assemble them into a finished production, complete with titles, graphics, and special effects, at the touch of a button!

DirectED PLUS is a title generator! Add brilliant titles in 12 different styles to your videos.

Choose from 64 colors. Superimpose titles over your video scenes or over color backgrounds.

Add titles in 12 styles and 64 colors.

DirectED PLUS is a graphics generator! Add your choice of 20 colorful, digital graphics, such as a birthday cake or wedding rings. Combine graphics and titles with any of 17 unique special effects to make your video scenes more personalized and exciting!

Combine titles with your choice of 20 colorful, digital graphics.

Combine titles, graphics, and special effects with your video footage.

DirectED PLUS is all you need! Use DirectED PLUS with the camcorder and VCR you already own. Virtually any VCR with a remote control is compatible with DirectED PLUS.

DirectED PLUS is surprisingly affordable! For a fraction of the price of a typical camcorder, DirectED PLUS turns your memories into polished, professional productions! Call 1-800-338-EDIT for the name of your nearest dealer.

Edit your best with **DirectED PLUS**

Some television screens are simulated. Specifications are subject to change without notice.©1989 Videonics, Inc.

▼ VIDEONICS

1370 DELL AVENUE • CAMPBELL, CA 95008 • 408-866-8300   *Call 1-800-338-EDIT for the name of your nearest dealer.*

Faster, more titles and graphics! DirectED owners can upgrade to DirectED PLUS for $49.95! Call 1-800-338-EDIT.

Add titles, animations, graphics, and sound effects to your videos without a computer

**NO COMPUTER NEEDED !
VHS and S-VHS !
CUSTOM TITLES ALSO !
OVER 1600 :**

ANIMATIONS
TITLES
GRAPHICS
SOUND EFFECTS

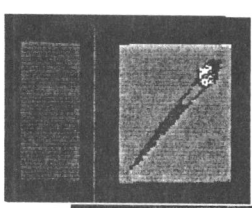

The **EASY ANIMATIONS** way

Easy Animations, Inc. has professionally recorded; over 1600 computer-generated, general purpose titles, animations, cartoons, dates, music, sound effects, 3d's, special effects, and graphics onto VHS and S-VHS videocassettes.

For professional or amateur videographers, who do not have access to expensive video computers and still insist on professional results in their work.

A video classroom will show you how to enhance your videos by adding or inserting our short video segments. Your finished video will be professional quality ! Post-production is fast with our exclusive 'on-screen' clock and companion 'index guide' which speeds you to the correct video segment. All you need is a camcorder and VCR or two VCRs. Enhancements may be done before shooting, on-location, or as post-production work. This will give your customers more options, thereby increasing your profits.

The low-cost way to add profit to your video business !

Volume One
THE ORIGINAL
over 400 titles

Children growing up
Animations and cartoons
Animated titles
Backgrounds
Birthdays
Dates, months and years
Vacations
Personal moments in sports
Weddings
Family chronicles
Holidays
Precious memories
Anniversaries
Days of the week
Special action graphics

And much, much more !

VHS - $ 29.95
S-VHS - $ 49.95

Volume Two
WEDDINGS
Professional Version
These faiths included:
Protestant
Episcopal
Baptist
Presbyterian
Methodist
Contemporary
Unitarian
Friends
Roman Catholic
Jewish
Interfaith
Orthodox Christian
Carpatho-Russian Orthodox
Muslim
Service of Renewal

VHS - $ 39.95
S-VHS - $ 59.95

Volume Three
500
MALE NAMES
To personalize weddings, birthdays, or any video.
Extra profit for you.
VHS - $ 29.95
S-VHS - $ 49.95

Volume Four
500
FEMALE NAMES
VHS - $ 29.95
S-VHS - $ 49.95

CUSTOM TITLES
(order custom workup kit first)

5 SLIDING TITLES
VHS - $ 39.95
S-VHS - $ 59.95

1 ANIMATED TITLE
Tumbling, rotating, or spinning
VHS - $ 39.95
S-VHS - $ 59.95

CUSTOM WORKUP KIT
Full credit toward custom titles.
Everything needed to select colors, fonts, sizes and transitions. Includes background 'blanks' for adding logos, art, pictures, etc.
$ 19.95

PROFESSIONAL SET-UP PACKAGE
Everything needed to add any title - OVER 1600 !

Includes :
Volume One
Volume Two
Volume Three
Volume Four
How-To Video
5 Transitional custom Titles
1 Animated Custom Title

VHS $ 199.95
S-VHS $ 319.95

SAVE

EASY ANIMATIONS, INC.
2121 Normal Park
Huntsville, Texas 77340
409-294-0481

To use VideoWare, either your VCR or camcorder must be VHS.

1-800-552-6402

MasterCard   VISA

Please add $ 1.55 per videocassette for shipping and handling

Guarantee
Return VideoWare prepaid, anytime, for a no-questions asked full refund.

Direct Re-Order Form

Name: _____

Address: _____

City: _____ State: ____ Zip: _____

Telephone Number: (___) _____

I would like to order _____ number of books @ $9.95 each. *(NOTE: If you paid less than $9.95 per book, enclose proof of purchase amount and we will only charge you that amount, plus $1.75 for UPS shipping per book.*

Send payment to: Video One Publications
3474 Dromedary Way, #1304
Las Vegas, NV 89115
(702) 643-2880

Want to write to the author? Here's your chance!

Comments? Questions? Recommendations?

Send your letter with SASE (self-addressed stamped envelope) to:
Video One Publications
Attn: John P. Johnston
3474 Dromedary Way, #1304
Las Vegas, NV 89115

or, if you want to spend the dime, call me at (702) 643-2880

LOOK FOR OUR NEW BOOKS:

EVERYTHING YOU ALWAYS WANTED TO KNOW ABOUT
VIDEO (But Nobody Had The Answers) $12.95 Plus Shipping
CANADA $19.95

EVERYTHING YOU ALWAYS WANTED TO KNOW ABOUT
VCRs (But Nobody Had The Answers) $9.95 Plus Shipping
CANADA $15.95

EVERYTHING YOU ALWAYS WANTED TO KNOW ABOUT
AUDIO (But Nobody Had The Answers)   Not yet available